Learn

# Eureka Math®
## Grade 4
## Modules 1 & 2

**Published by Great Minds®.**

Copyright © 2018 Great Minds®.

Printed in the U.S.A.

This book may be purchased from the publisher at eureka-math.org.

BAB   10   9   8   7   6   5   4   3   2

ISBN 978-1-64054-065-1

G4-M1-M2-L-05.2018

# Learn ◆ Practice ◆ Succeed

*Eureka Math*® student materials for *A Story of Units*® (K–5) are available in the *Learn, Practice, Succeed* trio. This series supports differentiation and remediation while keeping student materials organized and accessible. Educators will find that the *Learn, Practice,* and *Succeed* series also offers coherent—and therefore, more effective—resources for Response to Intervention (RTI), extra practice, and summer learning.

## Learn

*Eureka Math Learn* serves as a student's in-class companion where they show their thinking, share what they know, and watch their knowledge build every day. *Learn* assembles the daily classwork—Application Problems, Exit Tickets, Problem Sets, templates—in an easily stored and navigated volume.

## Practice

Each *Eureka Math* lesson begins with a series of energetic, joyous fluency activities, including those found in *Eureka Math Practice.* Students who are fluent in their math facts can master more material more deeply. With *Practice,* students build competence in newly acquired skills and reinforce previous learning in preparation for the next lesson.

Together, *Learn* and *Practice* provide all the print materials students will use for their core math instruction.

## Succeed

*Eureka Math Succeed* enables students to work individually toward mastery. These additional problem sets align lesson by lesson with classroom instruction, making them ideal for use as homework or extra practice. Each problem set is accompanied by a Homework Helper, a set of worked examples that illustrate how to solve similar problems.

Teachers and tutors can use *Succeed* books from prior grade levels as curriculum-consistent tools for filling gaps in foundational knowledge. Students will thrive and progress more quickly as familiar models facilitate connections to their current grade-level content.

# Students, families, and educators:

Thank you for being part of the *Eureka Math®* community, where we celebrate the joy, wonder, and thrill of mathematics.

In the *Eureka Math* classroom, new learning is activated through rich experiences and dialogue. The *Learn* book puts in each student's hands the prompts and problem sequences they need to express and consolidate their learning in class.

## What is in the Learn book?

**Application Problems:** Problem solving in a real-world context is a daily part of *Eureka Math*. Students build confidence and perseverance as they apply their knowledge in new and varied situations. The curriculum encourages students to use the RDW process—Read the problem, Draw to make sense of the problem, and Write an equation and a solution. Teachers facilitate as students share their work and explain their solution strategies to one another.

**Problem Sets:** A carefully sequenced Problem Set provides an in-class opportunity for independent work, with multiple entry points for differentiation. Teachers can use the Preparation and Customization process to select "Must Do" problems for each student. Some students will complete more problems than others; what is important is that all students have a 10-minute period to immediately exercise what they've learned, with light support from their teacher.

Students bring the Problem Set with them to the culminating point of each lesson: the Student Debrief. Here, students reflect with their peers and their teacher, articulating and consolidating what they wondered, noticed, and learned that day.

**Exit Tickets:** Students show their teacher what they know through their work on the daily Exit Ticket. This check for understanding provides the teacher with valuable real-time evidence of the efficacy of that day's instruction, giving critical insight into where to focus next.

**Templates:** From time to time, the Application Problem, Problem Set, or other classroom activity requires that students have their own copy of a picture, reusable model, or data set. Each of these templates is provided with the first lesson that requires it.

## Where can I learn more about Eureka Math resources?

The Great Minds® team is committed to supporting students, families, and educators with an ever-growing library of resources, available at eureka-math.org. The website also offers inspiring stories of success in the *Eureka Math* community. Share your insights and accomplishments with fellow users by becoming a *Eureka Math* Champion.

Best wishes for a year filled with aha moments!

*Jill Diniz*

Jill Diniz
Director of Mathematics
Great Minds

# The Read–Draw–Write Process

The *Eureka Math* curriculum supports students as they problem-solve by using a simple, repeatable process introduced by the teacher. The Read–Draw–Write (RDW) process calls for students to

1. Read the problem.
2. Draw and label.
3. Write an equation.
4. Write a word sentence (statement).

Educators are encouraged to scaffold the process by interjecting questions such as

- What do you see?
- Can you draw something?
- What conclusions can you make from your drawing?

The more students participate in reasoning through problems with this systematic, open approach, the more they internalize the thought process and apply it instinctively for years to come.

# Contents

## Module 4:  Place Value, Rounding, and Algorithms for Addition and Subtraction

## Module 2: Unit Conversions and Problem Solving with Metric Measurement

# Grade 4
# Module 1

Ben has a rectangular area 9 meters long and 6 meters wide. He wants a fence that will go around it as well as grass sod to cover it. How many meters of fence will he need? How many square meters of grass sod will he need to cover the entire area?

_____

_____

_____

_____

**Read**          **Draw**          **Write**

Name _____    Date _____

1.  Label the place value charts.  Fill in the blanks to make the following equations true.  Draw disks in the place value chart to show how you got your answer, using arrows to show any bundling.

    a.  $10 \times 3$ ones = ___30___ ones = _____

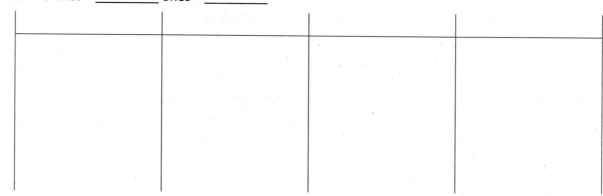

    b.  $10 \times 2$ tens = _____ tens = _____

    c.  4 hundreds $\times$ 10 = _____ hundreds = _____

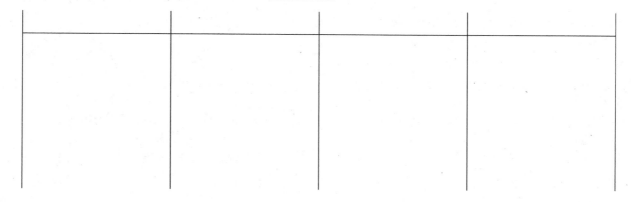

2.  Complete the following statements using your knowledge of place value:

    a.  10 times as many as 1 ten is _____ tens.

    b.  10 times as many as _____ tens is 30 tens or _____ hundreds.

    c.  _____ as 9 hundreds is 9 thousands.

    d.  _____ thousands is the same as 20 hundreds.

    Use pictures, numbers, or words to explain how you got your answer for Part (d).

3.  Matthew has 30 stamps in his collection.  Matthew's father has 10 times as many stamps as Matthew.
    How many stamps does Matthew's father have?  Use numbers or words to explain how you got your
    answer.

**Lesson 1:**      Interpret a multiplication equation as a comparison.

EUREKA
MATH

4.  Jane saved $800.  Her sister has 10 times as much money.  How much money does Jane's sister have?
    Use numbers or words to explain how you got your answer.

5.  Fill in the blanks to make the statements true.

    a.  2 times as much as 4 is _____.

    b.  10 times as much as 4 is _____.

    c.  500 is 10 times as much as _____.

    d.  6,000 is _____ as 600.

6.  Sarah is 9 years old.  Sarah's grandfather is 90 years old.  Sarah's grandfather is how many times as old as
    Sarah?

    Sarah's grandfather is _____ times as old as Sarah.

Name _____          Date _____

Use the disks in the place value chart below to complete the following problems:

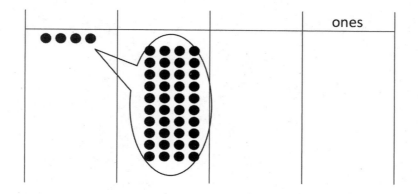

1.  Label the place value chart.


2.  Tell about the movement of the disks in the place value chart by filling in the blanks to make the following equation match the drawing in the place value chart:


        _____ × 10 = _____  = _____


3.  Write a statement about this place value chart using the words *10 times as many*.

unlabeled thousands place value chart

Amy is baking muffins.  Each baking tray can hold 6 muffins.

a.  If Amy bakes 4 trays of muffins, how many muffins will she have in all?

b.  The corner bakery produced 10 times as many muffins as Amy baked.  How many muffins did the bakery produce?

**Read**          **Draw**          **Write**

Lesson 2:        Recognize a digit represents 10 times the value of what it represents in the place to its right.

© 2018 Great Minds®. eureka-math.org

13

**Extension:** If the corner bakery packages the muffins in boxes of 100, how many boxes of 100 could they make?

**Read          Draw          Write**

**Lesson 2:**      Recognize a digit represents 10 times the value of what it represents in the place to its right.

EUREKA MATH

Name _____   Date _____

1. As you did during the lesson, label and represent the product or quotient by drawing disks on the place value chart.

   a.   10 × 2 thousands = _____ thousands = _____

   b.   10 × 3 ten thousands = _____ ten thousands = _____

   c.   4 thousands ÷ 10 = _____ hundreds ÷ 10 = _____

Lesson 2:   Recognize a digit represents 10 times the value of what it represents in the place to its right.

EUREKA MATH

© 2018 Great Minds®. eureka-math.org

15

2.  Solve for each expression by writing the solution in unit form and in standard form.

| Expression | Unit form | Standard Form |
|---|---|---|
| 10 × 6 tens | | |
| 7 hundreds × 10 | | |
| 3 thousands ÷ 10 | | |
| 6 ten thousands ÷ 10 | | |
| 10 × 4 thousands | | |

3.  Solve for each expression by writing the solution in unit form and in standard form.

| Expression | Unit form | Standard Form |
|---|---|---|
| (4 tens 3 ones) × 10 | | |
| (2 hundreds 3 tens) × 10 | | |
| (7 thousands 8 hundreds) × 10 | | |
| (6 thousands 4 tens) ÷ 10 | | |
| (4 ten thousands 3 tens) ÷ 10 | | |

4.  Explain how you solved 10 × 4 thousands.  Use a place value chart to support your explanation.

Lesson 2:    Recognize a digit represents 10 times the value of what it represents in
the place to its right.

EUREKA
MATH

5. Explain how you solved (4 ten thousands 3 tens) ÷ 10.  Use a place value chart to support your explanation.

6. Jacob saved 2 thousand dollar bills, 4 hundred dollar bills, and 6 ten dollar bills to buy a car.  The car costs 10 times as much as he has saved.  How much does the car cost?

7. Last year the apple orchard experienced a drought and did not produce many apples.  But this year, the apple orchard produced 45 thousand Granny Smith apples and 9 hundred Red Delicious apples, which is 10 times as many apples as last year.  How many apples did the orchard produce last year?

Lesson 2:    Recognize a digit represents 10 times the value of what it represents in
             the place to its right.

© 2018 Great Minds®. eureka-math.org

17

8. Planet Ruba has a population of 1 million aliens. Planet Zamba has 1 hundred thousand aliens.

   a. How many more aliens does Planet Ruba have than Planet Zamba?

   b. Write a sentence to compare the populations for each planet using the words *10 times as many*.

**Lesson 2:** Recognize a digit represents 10 times the value of what it represents in the place to its right.

© 2018 Great Minds®. eureka-math.org

Name _____     Date _____

1.  Fill in the blank to make a true number sentence.  Use standard form.

    a.  (4 ten thousands 6 hundreds) × 10 = _____

    b.  (8 thousands 2 tens) ÷ 10 = _____

2.  The Carson family saved up $39,580 for a new home.  The cost of their dream home is 10 times as much as they have saved.  How much does their dream home cost?

**Lesson 2:**    Recognize a digit represents 10 times the value of what it represents in the place to its right.

© 2018 Great Minds®. eureka-math.org

19

| | |
|---|---|
| | |
| | |
| | |
| | |
| | |
| | |
| | |

unlabeled millions place value chart

The school library has 10,600 books.  The town library has 10 times as many books.  How many books does the town library have?

_____

_____

_____

_____

**Read**          **Draw**          **Write**

EUREKA MATH

Lesson 3:    Name numbers within 1 million by building understanding of the place
value chart and placement of commas for naming base thousand units.

23

© 2018 Great Minds®. eureka-math.org

Name _____ Date _____

1. Rewrite the following numbers including commas where appropriate:

   a. 1234 _____   b. 12345 _____   c. 123456 _____

   d. 1234567 _____   e. 12345678901 _____

3. Solve each expression. Record your answer in standard form.

| Expression | Standard Form |
|---|---|
| 5 tens + 5 tens | |
| 3 hundreds + 7 hundreds | |
| 400 thousands + 600 thousands | |
| 8 thousands + 4 thousands | |

3. Represent each addend with place value disks in the place value chart. Show the composition of larger units from 10 smaller units. Write the sum in standard form.

   a. 4 thousands + 11 hundreds = _____

| millions | hundred thousands | ten thousands | thousands | hundreds | tens | ones |
|---|---|---|---|---|---|---|
| | | | | | | |

EUREKA MATH

Lesson 3: Name numbers within 1 million by building understanding of the place value chart and placement of commas for naming base thousand units.

© 2018 Great Minds®. eureka-math.org

25

b.   24 ten thousands + 11 thousands = _____

| millions | hundred thousands | ten thousands | thousands | hundreds | tens | ones |
|---|---|---|---|---|---|---|
|  |  |  |  |  |  |  |

4.   Use digits or disks on the place value chart to represent the following equations. Write the product in standard form.

a.   10 × 3 thousands = _____

How many thousands are in the answer? _____

| millions | hundred thousands | ten thousands | thousands | hundreds | tens | ones |
|---|---|---|---|---|---|---|
|  |  |  |  |  |  |  |

b.   (3 ten thousands 2 thousands) × 10 = _____

How many thousands are in the answer? _____

| millions | hundred thousands | ten thousands | thousands | hundreds | tens | ones |
|---|---|---|---|---|---|---|
|  |  |  |  |  |  |  |

Lesson 3:    Name numbers within 1 million by building understanding of the place value chart and placement of commas for naming base thousand units.

© 2018 Great Minds®. eureka-math.org

EUREKA MATH

c.  (32 thousands 1 hundred 4 ones) × 10 = _____

How many thousands are in your answer? _____

| millions | hundred thousands | ten thousands | thousands | hundreds | tens | ones |
|---|---|---|---|---|---|---|
|  |  |  |  |  |  |  |

5.  Lee and Gary visited South Korea.  They exchanged their dollars for South Korean bills.  Lee received 15 ten thousand South Korean bills.  Gary received 150 thousand bills.  Use disks or numbers on a place value chart to compare Lee's and Gary's money.

EUREKA MATH

Lesson 3:    Name numbers within 1 million by building understanding of the place value chart and placement of commas for naming base thousand units.

27

© 2018 Great Minds®. eureka-math.org

Name _____     Date _____

1.  In the spaces provided, write the following units in standard form. Be sure to place commas where appropriate.

   a.   9 thousands 3 hundreds 4 ones _____

   b.   6 ten thousands 2 thousands 7 hundreds 8 tens 9 ones _____

   c.   1 hundred thousand 8 thousands 9 hundreds 5 tens 3 ones _____

2.  Use digits or disks on the place value chart to write 26 thousands 13 hundreds.

| millions | hundred thousands | ten thousands | thousands | hundreds | tens | ones |
|---|---|---|---|---|---|---|
|  |  |  |  |  |  |  |

   How many thousands are in the number you have written? _____

**EUREKA MATH**     **Lesson 3:**    Name numbers within 1 million by building understanding of the place value chart and placement of commas for naming base thousand units.    **29**

© 2018 Great Minds®. eureka-math.org

There are about forty-one thousand Asian elephants and about four hundred seventy thousand African elephants left in the world.  About how many Asian and African elephants are left in total?

_____

_____

_____

_____

**Read          Draw          Write**

Lesson 4:    Read and write multi-digit numbers using base ten numerals, number names, and expanded form.

31

Name _____     Date _____

1.  a.   On the place value chart below, label the units, and represent the number 90,523.

| | | | | | | |
|---|---|---|---|---|---|---|
| | | | | | | |

    b.   Write the number in word form.

    c.   Write the number in expanded form.

2.  a.   On the place value chart below, label the units, and represent the number 905,203.

| | | | | | | |
|---|---|---|---|---|---|---|
| | | | | | | |

    b.   Write the number in word form.

    c.   Write the number in expanded form.

3. Complete the following chart:

| Standard Form | Word Form | Expanded Form |
|---|---|---|
| | two thousand, four hundred eighty | |
| | | 20,000 + 400 + 80 + 2 |
| | sixty-four thousand, one hundred six | |
| 604,016 | | |
| 960,060 | | |

4. Black rhinos are endangered, with only 4,400 left in the world. Timothy read that number as " four thousand, four hundred." His father read the number as "44 hundred." Who read the number correctly? Use pictures, numbers, or words to explain your answer.

Lesson 4:   Read and write multi-digit numbers using base ten numerals, number names, and expanded form.

EUREKA MATH®

Name _____     Date _____

1.  Use the place value chart below to complete the following:

| | | | | | | | |
|---|---|---|---|---|---|---|---|
| | | | | | | | |

   a.  Label the units on the chart.

   b.  Write the number 800,000 + 6,000 + 300 + 2 in the place value chart.

   c.  Write the number in word form.

2.  Write one hundred sixty thousand, five hundred eighty-two in expanded form.

Lesson 4:    Read and write multi-digit numbers using base ten numerals, number
             names, and expanded form.

© 2018 Great Minds®. eureka-math.org

35

Draw and label the units on the place value chart to hundred thousands.  Use each of the digits 9, 8, 7, 3, 1, and 0 once to create a number that is between 7 hundred thousands and 9 hundred thousands.  In word form, write the number you created.

**Extension:**  Create two more numbers following the same directions as above.

_____

_____

_____

_____

**Read**          **Draw**          **Write**

Name _____ Date _____

1.  Label the units in the place value chart.  Draw place value disks to represent each number in the place value chart.  Use <, >, or = to compare the two numbers.  Write the correct symbol in the circle.

    a.                              600,015  ( )  60,015

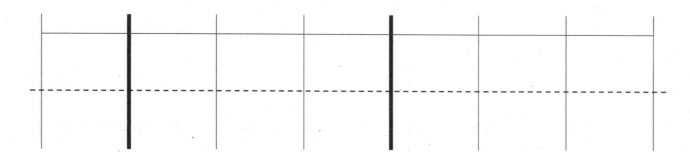

    b.                              409,004  ( )  440,002

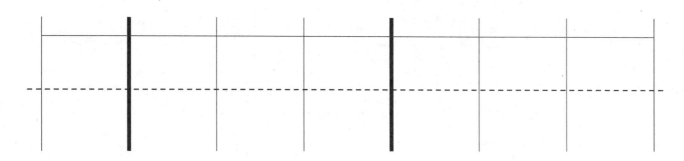

2.  Compare the two numbers by using the symbols <, >, and =.  Write the correct symbol in the circle.

    a.  342,001  ( )  94,981

    b.  500,000 + 80,000 + 9,000 + 100  ( )  five hundred eight thousand, nine hundred one

EUREKA MATH

Lesson 5:    Compare numbers based on meanings of the digits using >, <, or = to record the comparison.

© 2018 Great Minds®. eureka-math.org

39

c.   9 hundred thousands 8 thousands 9 hundreds 3 tens   ◯        908,930

d.   9 hundreds 5 ten thousands 9 ones   ◯   6 ten thousands 5 hundreds 9 ones

3.   Use the information in the chart below to list the height in feet of each mountain from least to greatest. Then, name the mountain that has the lowest elevation in feet.

| Name of Mountain | Elevation in Feet (ft) |
|---|---|
| Allen Mountain | 4,340 ft |
| Mount Marcy | 5,344 ft |
| Mount Haystack | 4,960 ft |
| Slide Mountain | 4,240 ft |

Compare numbers based on meanings of the digits using >, <, or = to record the comparison.

EUREKA
MATH

4. Arrange these numbers from least to greatest: 8,002   2,080   820   2,008   8,200

5. Arrange these numbers from greatest to least: 728,000   708,200   720,800   87,300

6. One astronomical unit, or 1 AU, is the approximate distance from Earth to the sun. The following are the approximate distances from Earth to nearby stars given in AUs:

Alpha Centauri is 275,725 AUs from Earth.
Proxima Centauri is 268,269 AUs from Earth.
Epsilon Eridani is 665,282 AUs from Earth.
Barnard's Star is 377,098 AUs from Earth.
Sirius is 542,774 AUs from Earth.

List the names of the stars and their distances in AUs in order from closest to farthest from Earth.

Lesson 5:    Compare numbers based on meanings of the digits using >, <, or = to
record the comparison.

© 2018 Great Minds®. eureka-math.org

41

Name _____  Date _____

1. Four friends played a game. The player with the most points wins. Use the information in the table below to order the number of points each player earned from least to greatest. Then, name the person who won the game.

| Player Name | Points Earned |
|---|---|
| Amy | 2,398 points |
| Bonnie | 2,976 points |
| Jeff | 2,709 points |
| Rick | 2,699 points |

2. Use each of the digits 5, 4, 3, 2, 1 exactly once to create two different five-digit numbers.

   a. Write each number on the line, and compare the two numbers by using the symbols < or >. Write the correct symbol in the circle.

   b. Use words to write a comparison statement for the problem above.

EUREKA MATH

Lesson 5:     Compare numbers based on meanings of the digits using >, <, or = to record the comparison.

© 2018 Great Minds®. eureka-math.org

43

| | |
|---|---|
| | |
| | |
| | |
| | |
| | |
| | |

unlabeled hundred thousands place value chart

Use the digits 5, 6, 8, 2, 4, and 1 to create two six-digit numbers. Be sure to use each of the digits within both numbers. Express the numbers in word form, and use a comparison symbol to show their relationship.

_____

_____

_____

_____

**Read          Draw          Write**

Name _____    Date _____

1.  Label the place value chart.  Use place value disks to find the sum or difference.  Write the answer in standard form on the line.

   a.  10,000 more than six hundred five thousand, four hundred seventy-two is _____.

   b.  100 thousand less than 400,000 + 80,000 + 1,000 + 30 + 6 is _____.

   c.  230,070 is _____ than 130,070.

2.  Lucy plays an online math game.  She scored 100,000 more points on Level 2 than on Level 3.  If she scored 349,867 points on Level 2, what was her score on Level 3?  Use pictures, words, or numbers to explain your thinking.

Lesson 6:    Find 1, 10, and 100 thousand more and less than a given number.

49

© 2018 Great Minds®. eureka-math.org

3. Fill in the blank for each equation.

   a. 10,000 + 40,060 = _____

   b. 21,195 − 10,000 = _____

   c. 999,000 + 1,000 = _____

   d. 129,231 − 100,000 = _____

   e. 122,000 = 22,000 + _____

   f. 38,018 = 39,018 − _____

4. Fill in the empty boxes to complete the patterns.

   a.

   | 150,010 | | 170,010 | | 190,010 | |
   |---|---|---|---|---|---|

   Explain in pictures, numbers, or words how you found your answers.

   b.

   | | 898,756 | 798,756 | | | 498,756 |
   |---|---|---|---|---|---|

   Explain in pictures, numbers, or words how you found your answers.

EUREKA MATH

c.

| 744,369 | 743,369 | | 741,369 | | |
|---|---|---|---|---|---|

Explain in pictures, numbers, or words how you found your answers.

d.

| | 118,910 | | | 88,910 | 78,910 |
|---|---|---|---|---|---|

Explain in pictures, numbers, or words how you found your answers.

Lesson 6:    Find 1, 10, and 100 thousand more and less than a given number.

51

© 2018 Great Minds®. eureka-math.org

Name _____     Date _____

1.  Fill in the empty boxes to complete the pattern.

| 468,235 | | | 471,235 | 472,235 | |
|---|---|---|---|---|---|

Explain in pictures, numbers, or words how you found your answers.

2.  Fill in the blank for each equation.

a.  1,000 + 56,879 = _____       b.  324,560 − 100,000 = _____

c.  456,080 − 10,000 = _____     d.  10,000 + 786,233 = _____

3.  The population of Rochester, NY, in the 2000 Census was 219,782.  The 2010 Census found that the population decreased by about 10,000.  About how many people lived in Rochester in 2010? Explain in pictures, numbers, or words how you found your answer.

According to their pedometers, Mrs. Alsup's class took a total of 42,619 steps on Tuesday. On Wednesday, they took ten thousand more steps than they did on Tuesday. On Thursday, they took one thousand fewer steps than they did on Wednesday. How many steps did Mrs. Alsup's class take on Thursday?

_____

_____

_____

_____

**Read**          **Draw**          **Write**

Name _____    Date _____

1.  Round to the nearest thousand.  Use the number line to model your thinking.

a.  6,700 ≈ _____

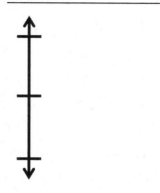

b.  9,340 ≈ _____

c.  16,401 ≈ _____

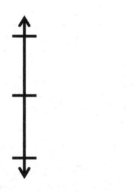

d.  39,545 ≈ _____

e.  399,499 ≈ _____

f.  840,007 ≈ _____

EUREKA
MATH®

Lesson 7:    Round multi-digit numbers to the thousands place using the vertical
number line.

© 2018 Great Minds®. eureka-math.org

57

2.  A pilot wanted to know about how many kilometers he flew on his last 3 flights.  From NYC to London, he flew 5,572 km.  Then, from London to Beijing, he flew 8,147 km.  Finally, he flew 10,996 km from Beijing back to NYC.  Round each number to the nearest thousand, and then find the sum of the rounded numbers to estimate about how many kilometers the pilot flew.

3.  Mrs. Smith's class is learning about healthy eating habits.  The students learned that the average child should consume about 12,000 calories each week.  Kerry consumed 12,748 calories last week.  Tyler consumed 11,702 calories last week.  Round to the nearest thousand to find who consumed closer to the recommended number of calories.  Use pictures, numbers, or words to explain.

4.  For the 2013-2014 school year, the cost of tuition at Cornell University was $43,000 when rounded to the nearest thousand.  What is the greatest possible amount the tuition could be?  What is the least possible amount the tuition could be?

**Lesson 7:**        Round multi-digit numbers to the thousands place using the vertical number line.

EUREKA
MATH®

Name _____    Date _____

1. Round to the nearest thousand. Use the number line to model your thinking.

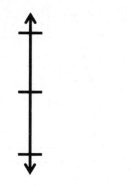  

   a. 7,621 ≈ _____    b. 12,502 ≈ _____    c. 324,087 ≈ _____

2. It takes 39,090 gallons of water to manufacture a new car. Sammy thinks that rounds up to about 40,000 gallons. Susie thinks it is about 39,000 gallons. Who rounded to the nearest thousand, Sammy or Susie? Use pictures, numbers, or words to explain.

EUREKA MATH®

Lesson 7:    Round multi-digit numbers to the thousands place using the vertical number line.

© 2018 Great Minds®. eureka-math.org

59

Jose's parents bought a used car, a new motorcycle, and a used snowmobile.  The car cost $8,999.  The motorcycle cost $9,690.  The snowmobile cost $4,419.  About how much money did they spend on the three items?

_____

_____

_____

_____

**Read          Draw          Write**

Name _____     Date _____

Complete each statement by rounding the number to the given place value.  Use the number line to show your work.

1.  a.  53,000 rounded to the nearest ten
        thousand is _____.

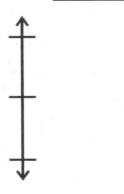

2.  a.  240,000 rounded to the nearest
        hundred thousand is _____.

    b.  42,708 rounded to the nearest ten
        thousand is _____.

    b.  449,019 rounded to the nearest
        hundred thousand is _____.

    c.  406,823 rounded to the nearest ten
        thousand is _____.

    c.  964,103 rounded to the nearest
        hundred thousand is _____.

Lesson 8:     Round multi-digit numbers to any place using the vertical number line.

© 2018 Great Minds®. eureka-math.org

63

3. 975,462 songs were downloaded in one day. Round this number to the nearest hundred thousand to estimate how many songs were downloaded in one day. Use a number line to show your work.

4. This number was rounded to the nearest ten thousand. List the possible digits that could go in the thousands place to make this statement correct. Use a number line to show your work.

$$13\_ ,644 \approx 130,000$$

5. Estimate the difference by rounding each number to the given place value.

$$712,350 - 342,802$$

a. Round to the nearest ten thousands.

b. Round to the nearest hundred thousands.

EUREKA MATH

Name _____    Date _____

1.  Round to the nearest ten thousand.  Use the number line to model your thinking.

    a.  35,124 ≈ _____          b.  981,657 ≈ _____

2.  Round to the nearest hundred thousand.  Use the number line to model your thinking.

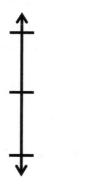

    a.  89,678 ≈ _____          b.  999,765 ≈ _____

3.  Estimate the sum by rounding each number to the nearest hundred thousand.

    257,098 + 548,765 ≈ _____

**EUREKA
MATH**®

**Lesson 8:**    Round multi-digit numbers to any place using the vertical number line.

65

© 2018 Great Minds®. eureka-math.org

34,123 people attended a basketball game. 28,310 people attended a football game. About how many more people attended the basketball game than the football game? Round to the nearest ten thousand to find the answer. Does your answer make sense? What might be a better way to compare attendance?

_____

_____

_____

_____

**Read**          **Draw**          **Write**

**Lesson 9:**     Use place value understanding to round multi-digit numbers to any place value.

© 2018 Great Minds®. eureka-math.org

Name _____    Date _____

1.  Round to the nearest thousand.

    a.  5,300 ≈ _____    b.  4,589 ≈ _____

    c.  42,099 ≈ _____    d.  801,504 ≈ _____

    e.  Explain how you found your answer for Part (d).

2.  Round to the nearest ten thousand.

    a.  26,000 ≈ _____    b.  34,920 ≈ _____

    c.  789,091 ≈ _____    d.  706,286 ≈ _____

    e.  Explain why two problems have the same answer.  Write another number that has the same answer when rounded to the nearest ten thousand.

3.  Round to the nearest hundred thousand.

    a.  840,000 ≈ _____    b.  850,471 ≈ _____

    c.  761,004 ≈ _____    d.  991,965 ≈ _____

    e.  Explain why two problems have the same answer.  Write another number that has the same answer when rounded to the nearest hundred thousand.

EUREKA MATH

Lesson 9:    Use place value understanding to round multi-digit numbers to any place value.

© 2018 Great Minds®. eureka-math.org

69

4.  Solve the following problems using pictures, numbers, or words.

    a.  The 2012 Super Bowl had an attendance of just 68,658 people. If the headline in the newspaper the next day read, " About 70,000 People Attend Super Bowl," how did the newspaper round to estimate the total number of people in attendance?

    b.  The 2011 Super Bowl had an attendance of 103,219 people. If the headline in the newspaper the next day read, " About 200,000 People Attend Super Bowl," is the newspaper's estimate reasonable? Use rounding to explain your answer.

    c.  According to the problems above, about how many more people attended the Super Bowl in 2011 than in 2012? Round each number to the largest place value before giving the estimated answer.

**Lesson 9:**     Use place value understanding to round multi-digit numbers to any place value.

EUREKA
MATH

Name _____     Date _____

1.  Round 765,903 to the given place value:

    Thousand                    _____

    Ten thousand                _____

    Hundred thousand            _____

2.  There are 16,850 Star coffee shops around the world.  Round the number of shops to the nearest thousand and ten thousand.  Which answer is more accurate?  Explain your thinking using pictures, numbers, or words.

The post office sold 204,789 stamps last week and 93,061 stamps this week.  About how many more stamps did the post office sell last week than this week?  Explain how you got your answer.

_____

_____

_____

_____

**Read**          **Draw**          **Write**

Lesson 10:     Use place value understanding to round multi-digit numbers to any
               place value using real world applications.

© 2018 Great Minds®. eureka-math.org

73

Name _____    Date _____

1.  Round 543,982 to the nearest

    a.  thousand: _____.

    b.  ten thousand: _____.

    c.  hundred thousand: _____.

2.  Complete each statement by rounding the number to the given place value.

    a.  2,841 rounded to the nearest hundred is _____.

    b.  32,851 rounded to the nearest hundred is _____.

    c.  132,891 rounded to the nearest hundred is _____.

    d.  6,299 rounded to the nearest thousand is _____.

    e.  36,599 rounded to the nearest thousand is _____.

    f.  100,699 rounded to the nearest thousand is _____.

    g.  40,984 rounded to the nearest ten thousand is _____.

    h.  54,984 rounded to the nearest ten thousand is _____.

    i.  997,010 rounded to the nearest ten thousand is _____.

    j.  360,034 rounded to the nearest hundred thousand is _____.

    k.  436,709 rounded to the nearest hundred thousand is _____.

    l.  852,442 rounded to the nearest hundred thousand is _____.

Lesson 10:    Use place value understanding to round multi-digit numbers to any
place value using real world applications.

© 2018 Great Minds®. eureka-math.org

75

3.  Empire Elementary School needs to purchase water bottles for field day. There are 2,142 students. Principal Vadar rounded to the nearest hundred to estimate how many water bottles to order. Will there be enough water bottles for everyone? Explain.

4.  Opening day at the New York State Fair in 2012 had an attendance of 46,753. Decide which place value to round 46,753 to if you were writing a newspaper article. Round the number and explain why it is an appropriate unit to round the attendance to.

5.  A jet airplane holds about 65,000 gallons of gas. It uses about 7,460 gallons when flying between New York City and Los Angeles. Round each number to the largest place value. Then, find about how many trips the plane can take between cities before running out of fuel.

Lesson 10: Use place value understanding to round multi-digit numbers to any place value using real world applications.

EUREKA MATH

Name _____   Date _____

1. There are 598,500 Apple employees in the United States.
   a. Round the number of employees to the given place value.

      thousand: _____

      ten thousand: _____

      hundred thousand: _____

   b. Explain why two of your answers are the same.

2. A company developed a student survey so that students could share their thoughts about school. In 2011, 78,234 students across the United States were administered the survey. In 2012, the company planned to administer the survey to 10 times as many students as were surveyed in 2011. About how many surveys should the company have printed in 2012? Explain how you found your answer.

Lesson 10:   Use place value understanding to round multi-digit numbers to any place value using real world applications.   77

© 2018 Great Minds®. eureka-math.org

Meredith kept track of the calories she consumed for three weeks.  The first week, she consumed 12,490 calories, the second week 14,295 calories, and the third week 11,116 calories.  About how many calories did Meredith consume altogether?  Which of these estimates will produce a more accurate answer: rounding to the nearest thousand or rounding to the nearest ten thousand?  Explain.

_____

_____

_____

_____

**Read**            **Draw**            **Write**

Lesson 11:    Use place value understanding to fluently add multi-digit whole numbers using the standard addition algorithm, and apply the algorithm to solve word problems using tape diagrams.

© 2018 Great Minds®. eureka-math.org

79

Name _____   Date _____

1.  Solve the addition problems below using the standard algorithm.

a.      6, 3 1 1
     +    2 6 8

b.      6, 3 1 1
     + 1, 2 6 8

c.      6, 3 1 4
     + 1, 2 6 8

d.      6, 3 1 4
     + 2, 4 9 3

e.      8, 3 1 4
     + 2, 4 9 3

f.      1 2, 3 7 8
     +    5, 4 6 3

g.      5 2, 0 9 8
     +    6, 0 4 8

h.      3 4, 6 9 8
     + 7 1, 8 4 0

i.      5 4 4, 8 1 1
     + 3 5 6, 4 4 5

j.      527 + 275 + 752

k.      38,193 + 6,376 + 241,457

Draw a tape diagram to represent each problem. Use numbers to solve, and write your answer as a statement.

2.  In September, Liberty Elementary School collected 32,537 cans for a fundraiser.  In October, they collected 207,492 cans. How many cans were collected during September and October?

3.  A baseball stadium sold some burgers.  2,806 were cheeseburgers.  1,679 burgers didn't have cheese. How many burgers did they sell in all?

4.  On Saturday night, 23,748 people attended the concert.  On Sunday, 7,570 more people attended the concert than on Saturday.  How many people attended the concert on Sunday?

**Lesson 11:**      Use place value understanding to fluently add multi-digit whole
numbers using the standard addition algorithm, and apply the
algorithm to solve word problems using tape diagrams.
© 2018 Great Minds®. eureka-math.org

EUREKA
MATH®

Name _____   Date _____

1. Solve the addition problems below using the standard algorithm.

   a.     2 3, 6 0 7
       +    2, 3 0 7
       _____

   b.     3, 9 4 8
       +     2 7 8
       _____

   c.   5,983 + 2,097

2. The office supply closet had 25,473 large paper clips, 13,648 medium paper clips, and 15,306 small paper clips. How many paper clips were in the closet?

| millions | hundred thousands | ten thousands | thousands | hundreds | tens | ones |
|---|---|---|---|---|---|---|
| | | | | | | |

millions place value char

Lesson 11:    Use place value understanding to fluently add multi-digit whole numbers using the standard addition algorithm, and apply the algorithm to solve word problems using tape diagrams.

© 2018 Great Minds®. eureka-math.org

85

The basketball team raised a total of $154,694 in September and $29,987 more in October than in September. How much money did they raise in October? Draw a tape diagram, and write your answer in a complete sentence.

_____

_____

_____

_____

**Read**          **Draw**          **Write**

Name _____    Date _____

Estimate and then solve each problem. Model the problem with a tape diagram. Explain if your answer is reasonable.

1.  For the bake sale, Connie baked 144 cookies. Esther baked 49 more cookies than Connie.

    a.  About how many cookies did Connie and Esther bake? Estimate by rounding each number to the nearest ten before adding.

    b.  Exactly how many cookies did Connie and Esther bake?

    c.  Is your answer reasonable? Compare your estimate from (a) to your answer from (b). Write a sentence to explain your reasoning.

Lesson 12:  Solve multi-step word problems using the standard addition algorithm modeled with tape diagrams, and assess the reasonableness of answers using rounding.

© 2018 Great Minds®. eureka-math.org

89

2.  Raffle tickets were sold for a school fundraiser to parents, teachers, and students.  563 tickets were sold to teachers.  888 more tickets were sold to students than to teachers.  904 tickets were sold to parents.

    a.  About how many tickets were sold to parents, teachers, and students?  Round each number to the nearest hundred to find your estimate.

    b.  Exactly how many tickets were sold to parents, teachers, and students?

    c.  Assess the reasonableness of your answer in (b).  Use your estimate from (a) to explain.

**Lesson 12:**     Solve multi-step word problems using the standard addition algorithm modeled with tape diagrams, and assess the reasonableness of answers using rounding.

EUREKA MATH

3.  From 2010 to 2011, the population of Queens increased by 16,075.  Brooklyn's population increased by 11,870 more than the population increase of Queens.

    a.  Estimate the total combined population increase of Queens and Brooklyn from 2010 to 2011. (Round the addends to estimate.)

    b.  Find the actual total combined population increase of Queens and Brooklyn from 2010 to 2011.

    c.  Assess the reasonableness of your answer in (b).  Use your estimate from (a) to explain.

**Lesson 12:**    Solve multi-step word problems using the standard addition algorithm modeled with tape diagrams, and assess the reasonableness of answers using rounding.

© 2018 Great Minds®. eureka-math.org

91

4. During National Recycling Month, Mr. Yardley's class spent 4 weeks collecting empty cans to recycle.

| Week | Number of Cans Collected |
|------|--------------------------|
| 1 | 10,827 |
| 2 | |
| 3 | 10,522 |
| 4 | 20,011 |

a. During Week 2, the class collected 1,256 more cans than they did during Week 1. Find the total number of cans Mr. Yardley's class collected in 4 weeks.

b. Assess the reasonableness of your answer in (a) by estimating the total number of cans collected.

Lesson 12:    Solve multi-step word problems using the standard addition algorithm
modeled with tape diagrams, and assess the reasonableness of
answers using rounding.

EUREKA
MATH

Name _____    Date _____

Model the problem with a tape diagram. Solve and write your answer as a statement.

In January, Scott earned $ 8,999.  In February, he earned $ 2,387 more than in January.  In March, Scott earned the same amount as in February.  How much did Scott earn altogether during those three months?  Is your answer reasonable?  Explain.

Jennifer texted 5,849 times in January.  In February, she texted 1,263 more times than in January. What was the total number of texts that Jennifer sent in the two months combined?  Explain how to know if the answer is reasonable.

_____

_____

_____

_____

**Read**            **Draw**            **Write**

Name _____  Date _____

1. Use the standard algorithm to solve the following subtraction problems.

    a.   7, 5 2 5
       − 3, 5 0 2

    b.   1 7, 5 2 5
       − 1 3, 5 0 2

    c.   6, 6 2 5
       − 4, 4 1 7

    d.   4, 6 2 5
       −    4 3 5

    e.   6, 5 0 0
       −    4 7 0

    f.   6, 0 2 5
       − 3, 5 0 2

    g.   2 3, 6 4 0
       − 1 4, 6 3 0

    h.   4 3 1, 9 2 5
       − 2 0 4, 8 1 5

    i.   2 1 9, 9 2 5
       − 1 2 1, 7 0 5

Draw a tape diagram to represent each problem.  Use numbers to solve, and write your answer as a statement. Check your answers.

2. What number must be added to 13,875 to result in a sum of 25,884?

EUREKA
MATH

Lesson 13:    Use place value understanding to decompose to smaller units once
using the standard subtraction algorithm, and apply the algorithm to
solve word problems using tape diagrams.

© 2018 Great Minds®. eureka-math.org

97

3.  Artist Michelangelo was born on March 6, 1475.  Author Mem Fox was born on March 6, 1946.  How many years after Michelangelo was born was Fox born?

4.  During the month of March, 68,025 pounds of king crab were caught.  If 15,614 pounds were caught in the first week of March, how many pounds were caught in the rest of the month?

5.  James bought a used car.  After driving exactly 9,050 miles, the odometer read 118,064 miles.  What was the odometer reading when James bought the car?

Lesson 13:    Use place value understanding to decompose to smaller units once using the standard subtraction algorithm, and apply the algorithm to solve word problems using tape diagrams.

EUREKA
MATH

Name _____     Date _____

1.  Use the standard algorithm to solve the following subtraction problems.

a.      8,512
     − 2,501

b.      18,042
     −  4,122

c.      8,072
     − 1,561

Draw a tape diagram to represent the following problem.  Use numbers to solve.  Write your answer as a statement.  Check your answer.

2.  What number must be added to 1,575 to result in a sum of 8,625?

In one year, the animal shelter bought 25,460 pounds of dog food.  That amount was 10 times the amount of cat food purchased in the month of July.  How much cat food was purchased in July?

**Extension:**  If the cats ate 1,462 pounds of the cat food, how much cat food was left?

_____

_____

_____

_____

**Read**          **Draw**          **Write**

**Lesson 14:**    Use place value understanding to decompose to smaller units up to three times using the standard subtraction algorithm, and apply the algorithm to solve word problems using tape diagrams.

© 2018 Great Minds®. eureka-math.org

101

Name _____   Date _____

1.  Use the standard algorithm to solve the following subtraction problems.

a.      2 , 4 6 0
      − 1 , 3 7 0

b.      2 , 4 6 0
      − 1 , 4 7 0

c.      9 7 , 6 8 4
      − 4 9 , 7 0 0

d.    1 2 4 , 3 0 6
      − 3 1 , 1 1 7

e.    1 2 4 , 3 0 6
      − 3 1 , 1 1 7

f.      9 7 , 6 8 4
      − 4 , 7 0 5

g.    1 2 4 , 0 0 6
    − 1 2 1 , 1 1 7

h.      9 7 , 6 8 4
      − 4 7 , 7 0 5

i.    1 2 4 , 0 6 0
      − 3 1 , 1 1 7

Draw a tape diagram to represent each problem.  Use numbers to solve, and write your answer as a statement.  Check your answers.

2.  There are 86,400 seconds in one day.  If Mr. Liegel is at work for 28,800 seconds a day, how many seconds a day is he away from work?

EUREKA MATH®

Lesson 14: Use place value understanding to decompose to smaller units up to three times using the standard subtraction algorithm, and apply the algorithm to solve word problems using tape diagrams.

© 2018 Great Minds®. eureka-math.org

103

3. A newspaper company delivered 240,900 newspapers before 6 a.m. on Sunday.  There were a total of 525,600 newspapers to deliver.  How many more newspapers needed to be delivered on Sunday?

4. A theater holds a total of 2,013 chairs.  197 chairs are in the VIP section.  How many chairs are not in the VIP section?

5. Chuck's mom spent $19,155 on a new car.  She had $30,064 in her bank account.  How much money does Chuck's mom have after buying the car?

Lesson 14:   Use place value understanding to decompose to smaller units up to three times using the standard subtraction algorithm, and apply the algorithm to solve word problems using tape diagrams.

© 2018 Great Minds®. eureka-math.org

Name _____     Date _____

Use the standard algorithm to solve the following subtraction problems.

1.        1 9, 3 5 0
         − 5, 7 6 1

2.    32,010 − 2,546

Draw a tape diagram to represent the following problem.  Use numbers to solve, and write your answer as a statement.  Check your answer.

3.  A doughnut shop sold 1,232 doughnuts in one day.  If they sold 876 doughnuts in the morning, how many doughnuts were sold during the rest of the day?

Lesson 14:   Use place value understanding to decompose to smaller units up to three times using the standard subtraction algorithm, and apply the algorithm to solve word problems using tape diagrams.

© 2018 Great Minds®. eureka-math.org

105

When the amusement park opened, the number on the counter at the gate read 928,614.  At the end of the day, the counter read 931,682.  How many people went through the gate that day?

_____

_____

_____

_____

**Read        Draw        Write**

**Lesson 15:** Use place value understanding to fluently decompose to smaller units multiple times in any place using the standard subtraction algorithm, and apply the algorithm to solve word problems using tape diagrams.

© 2018 Great Minds®. eureka-math.org

107

Name _____     Date _____

1.  Use the standard subtraction algorithm to solve the problems below.

a.      1 0 1, 6 6 0
      −     9 1, 6 8 0

b.      1 0 1, 6 6 0
      −        9, 9 8 0

c.      2 4 2, 5 6 1
      −     4 4, 7 0 2

d.      2 4 2, 5 6 1
      −     7 4, 9 8 7

e.      1, 0 0 0, 0 0 0
      −       5 9 2, 0 0 0

f.      1, 0 0 0, 0 0 0
      −       5 9 2, 5 0 0

g.        6 0 0, 6 5 8
      −   5 9 2, 5 6 9

h.        6 0 0, 0 0 0
      −   5 9 2, 5 6 9

**EUREKA MATH**

**Lesson 15:** Use place value understanding to fluently decompose to smaller units multiple times in any place using the standard subtraction algorithm, and apply the algorithm to solve word problems using tape diagrams.

© 2018 Great Minds®. eureka-math.org

109

Use tape diagrams and the standard algorithm to solve the problems below.  Check your answers.

2.  David is flying from Hong Kong to Buenos Aires.  The total flight distance is 11,472 miles.  If the plane has 7,793 miles left to travel, how far has it already traveled?

3.  Tank A holds 678,500 gallons of water.  Tank B holds 905,867 gallons of water.  How much less water does Tank A hold than Tank B?

4.  Mark had $25,081 in his bank account on Thursday.  On Friday, he added his paycheck to the bank account, and he then had $26,010 in the account.  What was the amount of Mark's paycheck?

**Lesson 15:**   Use place value understanding to fluently decompose to smaller units multiple times in any place using the standard subtraction algorithm, and apply the algorithm to solve word problems using tape diagrams.
© 2018 Great Minds®. eureka-math.org

EUREKA
MATH

Name _____    Date _____

Draw a tape diagram to model each problem and solve.

1.   956,204 – 780,169 = _____

2.   A construction company was building a stone wall on Main Street.  100,000 stones were delivered to the site.  On Monday, they used 15,631 stones.  How many stones remain for the rest of the week?  Write your answer as a statement.

 EUREKA MATH®

Lesson 15:    Use place value understanding to fluently decompose to smaller units
multiple times in any place using the standard subtraction algorithm,
and apply the algorithm to solve word problems using tape diagrams.
© 2018 Great Minds®. eureka-math.org

111

For the weekend basketball playoffs, a total of 61,941 tickets were sold. 29,855 tickets were sold for Saturday's games. The rest of the tickets were sold for Sunday's games. How many tickets were sold for Sunday's games?

_____

_____

_____

_____

**Read**       **Draw**       **Write**

Lesson 16:  Solve two-step word problems using the standard subtraction algorithm fluently modeled with tape diagrams, and assess the reasonableness of answers using rounding.

© 2018 Great Minds®. eureka-math.org

113

Name _____   Date _____

Estimate first, and then solve each problem.  Model the problem with a tape diagram.  Explain if your answer is reasonable.

1.  On Monday, a farmer sold 25,196 pounds of potatoes.  On Tuesday, he sold 18,023 pounds.
    On Wednesday, he sold some more potatoes.  In all, he sold 62,409 pounds of potatoes.

    a.  About how many pounds of potatoes did the farmer sell on Wednesday?  Estimate by rounding each value to the nearest thousand, and then compute.

    b.  Find the precise number of pounds of potatoes sold on Wednesday.

    c.  Is your precise answer reasonable?  Compare your estimate from (a) to your answer from (b).  Write a sentence to explain your reasoning.

2.  A gas station had two pumps.  Pump A dispensed 241,752 gallons.  Pump B dispensed 113,916 more gallons than Pump A.

   a.  About how many gallons did both pumps dispense?  Estimate by rounding each value to the nearest hundred thousand and then compute.

   b.  Exactly how many gallons did both pumps dispense?

   c.  Assess the reasonableness of your answer in (b).  Use your estimate from (a) to explain.

Lesson 16:     Solve two-step word problems using the standard subtraction algorithm fluently modeled with tape diagrams, and assess the reasonableness of answers using rounding.
© 2018 Great Minds®. eureka-math.org

EUREKA MATH®

3. Martin's car had 86,456 miles on it. Of that distance, Martin's wife drove 24,901 miles, and his son drove 7,997 miles. Martin drove the rest.

a. About how many miles did Martin drive? Round each value to estimate.

b. Exactly how many miles did Martin drive?

c. Assess the reasonableness of your answer in (b). Use your estimate from (a) to explain.

4. A class read 3,452 pages the first week and 4,090 more pages in the second week than in the first week. How many pages had they read by the end of the second week? Is your answer reasonable? Explain how you know using estimation.

5. A cargo plane weighed 500,000 pounds. After the first load was taken off, the airplane weighed 437,981 pounds. Then 16,478 more pounds were taken off. What was the total number of pounds of cargo removed from the plane? Is your answer reasonable? Explain.

Lesson 16:    Solve two-step word problems using the standard subtraction algorithm fluently modeled with tape diagrams, and assess the reasonableness of answers using rounding.

© 2018 Great Minds®. eureka-math.org

Name _____     Date _____

Quarterback Brett Favre passed for 71,838 yards between the years 1991 and 2011. His all-time high was 4,413 passing yards in one year. In his second highest year, he threw 4,212 passing yards.

1. About how many passing yards did he throw in the remaining years? Estimate by rounding each value to the nearest thousand and then compute.

2. Exactly how many passing yards did he throw in the remaining years?

3. Assess the reasonableness of your answer in (b). Use your estimate from (a) to explain.

A bakery used 12,674 kg of flour.  Of that, 1,802 kg was whole wheat and 888 kg was rice flour.  The rest was all-purpose flour.  How much all-purpose flour did they use?  Solve and check the reasonableness of your answer.

_____

_____

_____

_____

**Read**          **Draw**          **Write**

**Lesson 17:**     Solve *additive compare* word problems modeled with tape diagrams.

© 2018 Great Minds®. eureka-math.org

121

Name _____    Date _____

Draw a tape diagram to represent each problem.  Use numbers to solve, and write your answer as a statement.

1.   Sean's school raised $32,587.  Leslie's school raised $18,749.  How much more money did Sean's school raise?

2.   At a parade, 97,853 people sat in bleachers, and 388,547 people stood along the street.  How many fewer people were in the bleachers than standing on the street?

EUREKA
MATH

Lesson 17:    Solve *additive compare* word problems modeled with tape diagrams.

© 2018 Great Minds®. eureka-math.org

123

3. A pair of hippos weighs 5,201 kilograms together. The female weighs 2,038 kilograms. How much more does the male weigh than the female?

4. A copper wire was 240 meters long. After 60 meters was cut off, it was double the length of a steel wire. How much longer was the copper wire than the steel wire at first?

Lesson 17: Solve *additive compare* word problems modeled with tape diagrams.

EUREKA MATH

Name _____     Date _____

Draw a tape diagram to represent each problem.  Use numbers to solve, and write your answer as a statement.

A mixture of 2 chemicals measures 1,034 milliliters.  It contains some of Chemical A and 755 milliliters of Chemical B.  How much less of Chemical A than Chemical B is in the mixture?

In all, 30,436 people went skiing in February and January. 16,009 went skiing in February. How many fewer people went skiing in January than in February?

_____

_____

_____

_____

**Read**          **Draw**          **Write**

**Lesson 18:**   Solve multi-step word problems modeled with tape diagrams, and
assess the reasonableness of answers using rounding.

© 2018 Great Minds®. eureka-math.org

127

Name _____     Date _____

Draw a tape diagram to represent each problem. Use numbers to solve, and write your answer as a statement.

1. In one year, the factory used 11,650 meters of cotton, 4,950 fewer meters of silk than cotton, and 3,500 fewer meters of wool than silk. How many meters in all were used of the three fabrics?

2. The shop sold 12,789 chocolate and 9,324 cookie dough cones. It sold 1,078 more peanut butter cones than cookie dough cones and 999 more vanilla cones than chocolate cones. What was the total number of ice cream cones sold?

Lesson 18:    Solve multi-step word problems modeled with tape diagrams, and assess the reasonableness of answers using rounding.      **129**

© 2018 Great Minds®. eureka-math.org

3.  In the first week of June, a restaurant sold 10,345 omelets.  In the second week, 1,096 fewer omelets were sold than in the first week.  In the third week, 2 thousand more omelets were sold than in the first week.  In the fourth week, 2 thousand fewer omelets were sold than in the first week.  How many omelets were sold in all in June?

Lesson 18:    Solve multi-step word problems modeled with tape diagrams, and
assess the reasonableness of answers using rounding.

© 2018 Great Minds®. eureka-math.org

EUREKA
MATH

Name _____     Date _____

Draw a tape diagram to represent the problem.  Use numbers to solve, and write your answer as a statement.

Park A covers an area of 4,926 square kilometers.  It is 1,845 square kilometers larger than Park B.
Park C is 4,006 square kilometers larger than Park A.

1.  What is the area of all three parks?

2.  Assess the reasonableness of your answer.

**Lesson 18:**     Solve multi-step word problems modeled with tape diagrams, and
                   assess the reasonableness of answers using rounding.                                131

© 2018 Great Minds®. eureka-math.org

For Jordan to get to his grandparents' house, he has to travel through Albany and Plattsburgh. From Jordan's house to Albany is 189 miles. From Albany to Plattsburgh is 161 miles. If the total distance of the trip is 508 miles, how far from Plattsburgh do Jordan's grandparents live?

_____

_____

_____

_____

**Read**          **Draw**          **Write**

EUREKA MATH

Lesson 19:   Create and solve multi-step word problems from given tape diagrams and equations.

© 2018 Great Minds®. eureka-math.org

133

Name _____   Date _____

Using the diagrams below, create your own word problem.  Solve for the value of the variable.

1.

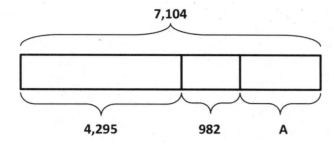

2.

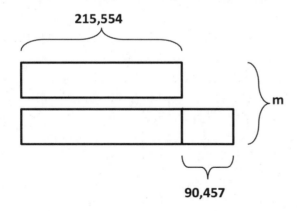

Lesson 19:   Create and solve multi-step word problems from given tape diagrams          135
             and equations.

© 2018 Great Minds®. eureka-math.org

3.

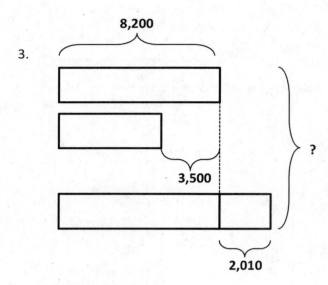

8,200

3,500

2,010

?

---

4. Draw a tape diagram to model the following equation. Create a word problem. Solve for the value of the variable.

$$26,854 = 17,729 + 3,731 + A$$

**Lesson 19:** Create and solve multi-step word problems from given tape diagrams and equations.

EUREKA
MATH

Name _____    Date _____

Using the diagram below, create your own word problem.  Solve for the value of the variable.

1.

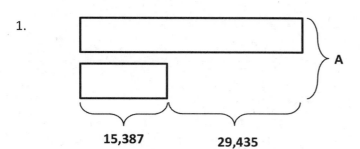

2.    Using the equation below, draw a tape diagram and create your own word problem.  Solve for the value of the variable.

$$248{,}798 = 113{,}205 + A + 99{,}937$$

Lesson 19:    Create and solve multi-step word problems from given tape diagrams and equations.

137

# Grade 4
# Module 2

Martha, George, and Elizabeth sprint a combined distance of 10,000 meters.  Martha sprints 3,206 meters.  George sprints 2,094 meters.  How far does Elizabeth sprint? Solve using an algorithm or a simplifying strategy.

_____

_____

_____

_____

**Read**          **Draw**          **Write**

Lesson 1:    Express metric length measurements in terms of a smaller unit;
model and solve addition and subtraction word problems involving
metric length.

© 2018 Great Minds®. eureka-math.org

141

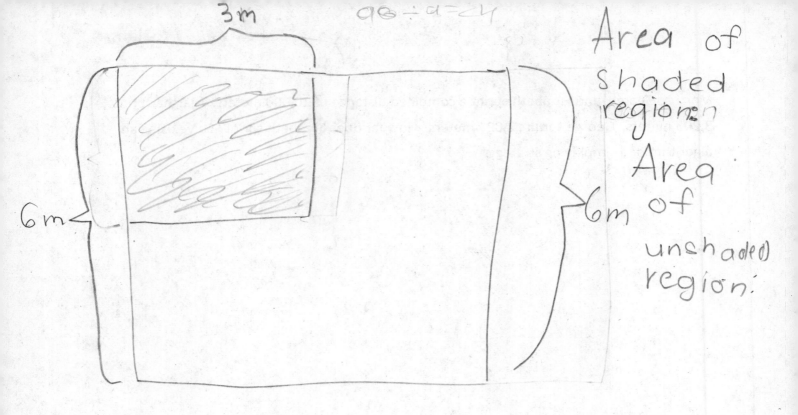

96 ÷ 4 = 24

3m

Area of shaded region:

6m

6m of unshaded region:

96 pencils were delivered today at school.
There was 4 teachers, each teacher gets
a amount of pencils in a box. How
many does each teacher recive? Show RDW

96 ÷ 4 = 24

24 ÷ 4 = 6

Name _____ Date _____

1. Convert the measurements.

   a. 1 km = ___1000___ m

   b. 4 km = ___4000___ m

   c. 7 km = ___7000___ m

   d. ___18___ km = 18,000 m

   e. 1 m = ___1000___ cm

   f. 3 m = ___3000___ cm

   g. 80 m = ___80,000___ cm

   h. ___12___ m = 12,000 cm

2. Convert the measurements.

   a. 3 km 312 m = ___3312___ m

   b. 13 km 27 m = ___13,027___ m

   c. 915 km 8 m = ___91508___ m

   d. 3 m 56 cm = _____ cm

   e. 14 m 8 cm = _____ cm

   f. 120 m 46 cm = _____ cm

3. Solve.

   a. 4 km − 280 m

   b. 1 m 15 cm − 34 cm

   c. Express your answer in the smaller unit:
      1 km 431 m + 13 km 169 m

   d. Express your answer in the smaller unit:
      231 m 31 cm − 14 m 48 cm

   e. 67 km 230 m + 11 km 879 m

   f. 67 km 230 m − 11 km 879 m

Use a tape diagram to model each problem.  Solve using a simplifying strategy or an algorithm, and write your answer as a statement.

4.  The length of Carter's driveway is 12 m 38 cm.  His neighbor's driveway is 4 m 99 cm longer.  How long is his neighbor's driveway?

5.  Enya walked 2 km 309 m from school to the store.  Then, she walked from the store to her home.  If she walked a total of 5 km, how far was it from the store to her home?

6.  Rachael has a rope 5 m 32 cm long that she cut into two pieces.  One piece is 249 cm long.  How many centimeters long is the other piece of rope?

7.  Jason rode his bike 529 fewer meters than Allison.  Jason rode 1 km 850 m.  How many meters did Allison ride?

Lesson 1:        Express metric length measurements in terms of a smaller unit; model and solve addition and subtraction word problems involving metric length.
© 2018 Great Minds®. eureka-math.org

EUREKA
MATH

Name _____ Date _____

1. Complete the conversion table.

| Distance | |
|---|---|
| 71 km | _____ m |
| _____ km | 30,000 m |
| 81 m | _____ cm |
| _____ m | 400 cm |

2. 13 km 20 m = _____ m

3. 401 km 101 m – 34 km 153 m = _____

4. Gabe built a toy tower that measured 1 m 78 cm. After building some more, he measured it, and it was 82 cm taller. How tall is his tower now? Draw a tape diagram to model this problem. Use a simplifying strategy or an algorithm to solve, and write your answer as a statement.

EUREKA
MATH

Lesson 1:     Express metric length measurements in terms of a smaller unit;
              model and solve addition and subtraction word problems involving
              metric length.
© 2018 Great Minds®. eureka-math.org

145

The distance from school to Zoie's house is 3 kilometers 469 meters. Camie's house is 4 kilometers 301 meters farther away from Zoie's. How far is it from Camie's house to school? Solve using an algorithm or a simplifying strategy.

_____

_____

_____

_____

**Read**          **Draw**          **Write**

Name _____  Date _____

1. Complete the conversion table.

| Mass | |
|---|---|
| **kg** | **g** |
| 1 | 1,000 |
| 3 | |
| | 4,000 |
| 17 | |
| | 20,000 |
| 300 | |

2. Convert the measurements.

a.     1 kg 500 g  =  _____ g

b.     3 kg 715 g  =  _____ g

c.     17 kg 84 g  =  _____ g

d.     25 kg 9 g  =  _____ g

e.   _____ kg _____ g  =      7,481 g

f.     210 kg 90 g  =  _____ g

3. Solve.
   a.  3,715 g – 1,500 g

   b. 1 kg – 237 g

   c. Express the answer in the smaller unit:
      25 kg 9 g + 24 kg  991 g

   d. Express the answer in the smaller unit:
      27 kg 650 g – 20 kg 990 g

   e. Express the answer in mixed units:
      14 kg 505 g – 4,288 g

   f. Express the answer in mixed units:
      5 kg 658 g + 57,481 g

EUREKA MATH®

Lesson 2:   Express metric mass measurements in terms of a smaller unit; model and solve addition and subtraction word problems involving metric mass.

© 2018 Great Minds®. eureka-math.org

149

Use a tape diagram to model each problem.  Solve using a simplifying strategy or an algorithm, and write your answer as a statement.

4.  One package weighs 2 kilograms 485 grams.  Another package weighs 5 kilograms 959 grams.  What is the total weight of the two packages?

5.  Together, a pineapple and a watermelon weigh 6 kilograms 230 grams.  If the pineapple weighs 1 kilogram 255 grams, how much does the watermelon weigh?

6.  Javier's dog weighs 3,902 grams more than Bradley's dog.  Bradley's dog weighs 24 kilograms 175 grams.  How much does Javier's dog weigh?

7.  The table to the right shows the weight of three Grade 4 students.  How much heavier is Isabel than the lightest student?

| Student | Weight |
| --- | --- |
| Isabel | 35 kg |
| Irene | 29 kg 38 g |
| Sue | 29,238 g |

Lesson 2:   Express metric mass measurements in terms of a smaller unit; model and solve addition and subtraction word problems involving metric mass.

© 2018 Great Minds®. eureka-math.org

EUREKA
MATH®

Name_____     Date _____

1. Convert the measurements.

   a.   21 kg 415 g = _____ g                    b.   2 kg 91 g = _____ g

   c.   87 kg 17 g = _____ g                      d.   _____ kg _____ g = 96,020 g

Use a tape diagram to model the following problem.  Solve using a simplifying strategy or an algorithm, and write your answer as a statement.

2. The table to the right shows the weight of three dogs. How much more does the Great Dane weigh than the Chihuahua?

| Dog | Weight |
|---|---|
| Great Dane | 59 kg |
| Golden Retriever | 32 kg 48 g |
| Chihuahua | 1,329 g |

EUREKA
MATH®

Lesson 2:     Express metric mass measurements in terms of a smaller unit; model and solve addition and subtraction word problems involving metric mass.

© 2018 Great Minds®. eureka-math.org

151

A liter of water weighs 1 kilogram.  The Lee family took 3 liters of water with them on a hike.  At the end of the hike, they had 290 grams of water left.  How much water did they drink?  Draw a tape diagram, and solve using an algorithm or a simplifying strategy.

_____

_____

_____

_____

**Read**        **Draw**        **Write**

EUREKA MATH®    Lesson 3:    Express metric capacity measurements in terms of a smaller unit; model and solve addition and subtraction word problems involving metric capacity.    153

© 2018 Great Minds®. eureka-math.org

Name _____    Date _____

1.  Complete the conversion table.

| Liquid Capacity | |
| --- | --- |
| L | mL |
| 1 | 1,000 |
| 5 | |
| 38 | |
| | 49,000 |
| 54 | |
| | 92,000 |

2.  Convert the measurements.

a.     2 L 500 mL   =   _____ mL

b.     70 L 850 mL   =   _____ mL

c.     33 L 15 mL   =   _____ mL

d.     2 L 8 mL   =   _____ mL

e.     3,812 mL   =   _____ L _____ mL

f.     86,003 mL   =   _____ L _____ mL

3.  Solve.

a.  1,760 mL + 40 L

b.  7 L – 3,400 mL

c.  Express the answer in the smaller unit:
    25 L 478 mL + 3 L 812 mL

d.  Express the answer in the smaller unit:
    21 L – 2 L 8 mL

e.  Express the answer in mixed units:
    7 L 425 mL – 547 mL

f.  Express the answer in mixed units:
    31 L 433 mL – 12 L 876 mL

Lesson 3:    Express metric capacity measurements in terms of a smaller unit;
             model and solve addition and subtraction word problems involving
             metric capacity.
© 2018 Great Minds®. eureka-math.org

Use a tape diagram to model each problem. Solve using a simplifying strategy or an algorithm, and write your answer as a statement.

4. To make fruit punch, John's mother combined 3,500 milliliters of tropical drink, 3 liters 95 milliliters of ginger ale, and 1 liter 600 milliliters of pineapple juice.

   a. Order the quantity of each drink from least to greatest.

   b. How much punch did John's mother make?

5. A family drank 1 liter 210 milliliters of milk at breakfast. If there were 3 liters of milk before breakfast, how much milk is left?

6. Petra's fish tank contains 9 liters 578 milliliters of water. If the capacity of the tank is 12 liters 455 milliliters of water, how many more milliliters of water does she need to fill the tank?

**Lesson 3:**      Express metric capacity measurements in terms of a smaller unit; model and solve addition and subtraction word problems involving metric capacity.

© 2018 Great Minds®. eureka-math.org

Name _____ Date _____

1. Convert the measurements.

   a. 6 L 127 mL = _____ mL

   b. 706 L 220 mL = _____ mL

   c. 12 L 9 mL = _____ mL

   d. _____ L _____ mL = 906,010 mL

2. Solve.

   81 L 603 mL – 22 L 489 mL

Use a tape diagram to model the following problem. Solve using a simplifying strategy or an algorithm, and write your answer as a statement.

3. The Smith's hot tub has a capacity of 1,458 liters. Mrs. Smith put 487 liters 750 milliliters of water in the tub. How much water needs to be added to fill the hot tub completely?

Lesson 3: Express metric capacity measurements in terms of a smaller unit; model and solve addition and subtraction word problems involving metric capacity.

© 2018 Great Minds®. eureka-math.org

157

Adam poured 1 liter 460 milliliters of water into a beaker.  Over three days, some of the water evaporated.  On the fourth day, 979 milliliters of water remained in the beaker.  How much water evaporated?

**Read**          **Draw**          **Write**

**EUREKA MATH**®

**Lesson 4:**     Know and relate metric units to place value units in order to express
                  measurements in different units.

© 2018 Great Minds®. eureka-math.org

159

Name _____     Date _____

1. Complete the table.

| Smaller Unit | Larger Unit | How Many Times as Large as? |
|---|---|---|
| one | hundred | 100 |
| centimeter | | 100 |
| one | thousand | 1,000 |
| gram | | 1,000 |
| meter | kilometer | |
| milliliter | | 1,000 |
| centimeter | kilometer | |

2. Fill in the units in word form.

   a. 429 is 4 hundreds 29 _____.

   b. 429 cm is 4 _____ 29 cm.

   c. 2,456 is 2 _____ 456 ones.

   d. 2,456 m is 2 _____ 456 m.

   e. 13,709 is 13 _____ 709 ones.

   f. 13,709 g is 13 kg 709 _____.

3. Fill in the unknown number.

   a. _____ is 456 thousands 829 ones.

   b. _____ mL is 456 L 829 mL.

Lesson 4:    Know and relate metric units to place value units in order to express measurements in different units.

© 2018 Great Minds®. eureka-math.org

161

4. Use words, equations, or pictures to show and explain how metric units are like and unlike place value units.

5. Compare using >, <, or =.

   a.  893,503 mL        ◯        89 L 353 mL

   b.  410 km 3 m        ◯        4,103 m

   c.  5,339 m           ◯        533,900 cm

6. Place the following measurements on the number line:

         2 km 415 m         2,379 m         2 km 305 m      245,500 cm

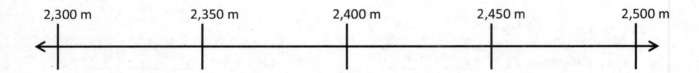

7. Place the following measurements on the number line:

         2 kg 900 g        3,500 g      1 kg 500 g      2,900 g      750 g

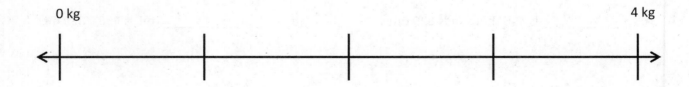

Know and relate metric units to place value units in order to express measurements in different units.

EUREKA MATH

© 2018 Great Minds®. eureka-math.org

Name _____    Date _____

1.  Fill in the unknown unit in word form.

    a.  8,135 is 8 _____ 135 ones.        b.  8,135 g is 8 _____ 135 g.

2.  _____ mL is equal to 342 L 645 mL.

3.  Compare using >, <, or =.

    a.  23 km 40 m    ◯    2,340 m

    b.  13,798 mL    ◯    137 L 980 mL

    c.  5,607 m    ◯    560,701 cm

4.  Place the following measurements on the number line:

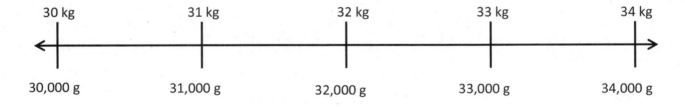

EUREKA
MATH®

Lesson 4:

Know and relate metric units to place value units in order to express
measurements in different units.

© 2018 Great Minds®. eureka-math.org

163

| | |
|---|---|
| | |
| | |
| | |
| | |
| | |
| | |

unlabeled hundred thousands place value chart

**Lesson 4:** Know and relate metric units to place value units in order to express measurements in different units.

165

© 2018 Great Minds®. eureka-math.org

Name _____    Date _____

Model each problem with a tape diagram.  Solve and answer with a statement.

1.  The potatoes Beth bought weighed 3 kilograms 420 grams.  Her onions weighed 1,050 grams less than the potatoes.  How much did the potatoes and onions weigh together?

2.  Adele let out 18 meters 46 centimeters of string to fly her kite.  She then let out 13 meters 78 centimeters more before reeling back in 590 centimeters.  How long was her string after reeling it in?

3.  Shyan's barrel contained 6 liters 775 milliliters of paint.  She poured in 1 liter 118 milliliters more.  The first day, Shyan used 2 liters 125 milliliters of the paint.  At the end of the second day, there were 1,769 milliliters of paint remaining in the barrel.  How much paint did Shyan use on the second day?

Lesson 5:     Use addition and subtraction to solve multi-step word problems
              involving length, mass, and capacity.

© 2018 Great Minds®. eureka-math.org

167

4. On Thursday, the pizzeria used 2 kilograms 180 grams less flour than they used on Friday. On Friday, they used 12 kilograms 240 grams. On Saturday, they used 1,888 grams more than on Friday. What was the total amount of flour used over the three days?

5. The gas tank in Zachary's car has a capacity of 60 liters. He adds 23 liters 825 milliliters of gas to the tank, which already has 2,050 milliliters of gas. How much more gas can Zachary add to the gas tank?

6. A giraffe is 5 meters 20 centimeters tall. An elephant is 1 meter 77 centimeters shorter than the giraffe. A rhinoceros is 1 meter 58 centimeters shorter than the elephant. How tall is the rhinoceros?

**Lesson 5:** Use addition and subtraction to solve multi-step word problems involving length, mass, and capacity.

EUREKA MATH

Name _____     Date _____

Model each problem with a tape diagram.  Solve and answer with a statement.

1.  Jeff places a pineapple with a mass of 890 grams on a balance scale.
    He balances the scale by placing two oranges, an apple, and a lemon
    on the other side.  Each orange weighs 280 grams.  The lemon weighs
    195 grams less than each orange.  What is the mass of the apple?

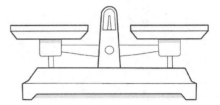

2.  Brian is 1 meter 87 centimeters tall.  Bonnie is 58 centimeters shorter than Brian.  Betina is
    26 centimeters taller than Bonnie.  How tall is Betina?

# Credits

Great Minds® has made every effort to obtain permission for the reprinting of all copyrighted material. If any owner of copyrighted material is not acknowledged herein, please contact Great Minds for proper acknowledgment in all future editions and reprints of this module.